ENERGY SECTOR STANDARD OF THE PEOPLE'S REPUBLIC OF CHINA

中华人民共和国能源行业标准

Specification for Preparation of Pre-feasibility Study Report for Photovoltaic Power Projects

光伏发电工程预可行性研究报告编制规程

NB/T 32044-2018

Chief Development Department: China Renewable Energy Engineering Institute
Approval Department: National Energy Administration of the People's Republic of China
Implementation Date: October 1, 2018

China Water & Power Press
中国水利水电出版社
Beijing 2024

All rights reserved. No part of this publication may be reproduced, stored in a retrieval system, or transmitted in any form or by any means—electronic, mechanical, photocopying, recording or otherwise, without prior written permission of the publisher.

图书在版编目（CIP）数据

光伏发电工程预可行性研究报告编制规程：NB/T 32044-2018 = Specification for Preparation of Pre-feasibility Study Report for Photovoltaic Power Projects(NB/T 32044-2018)：英文 / 国家能源局发布. 北京：中国水利水电出版社，2024. 10. -- ISBN 978-7-5226-2766-3

Ⅰ. TM615-65

中国国家版本馆CIP数据核字第202426KT89号

ENERGY SECTOR STANDARD
OF THE PEOPLE'S REPUBLIC OF CHINA
中华人民共和国能源行业标准

Specification for Preparation of Pre-feasibility Study Report
for Photovoltaic Power Projects
光伏发电工程预可行性研究报告编制规程
NB/T 32044-2018
（英文版）

Issued by National Energy Administration of the People's Republic of China
国家能源局　发布
Translation organized by China Renewable Energy Engineering Institute
水电水利规划设计总院　组织翻译
Published by China Water & Power Press
中国水利水电出版社　出版发行
　　Tel: (+ 86 10) 68545888　68545874
　　sales@mwr.gov.cn
　　Account name: China Water & Power Press
　　Address: No.1, Yuyuantan Nanlu, Haidian District, Beijing 100038, China
　　http://www.waterpub.com.cn
中国水利水电出版社微机排版中心　排版
北京中献拓方科技发展有限公司　印刷
184mm×260mm　16开本　3.25印张　103千字
2024年10月第1版　2024年10月第1次印刷

Price（定价）：￥400.00

Introduction

This English version is one of China's energy sector standard series in English. Its translation was organized by China Renewable Energy Engineering Institute authorized by National Energy Administration of the People's Republic of China in compliance with relevant procedures and stipulations. This English version was issued by National Energy Administration of the People's Republic of China in Announcement [2022] No. 4 dated May13, 2022.

This version was translated from the Chinese Standard NB/T 32044-2018, *Specification for Preparation of Pre-feasibility Study Report for Photovoltaic Power Projects*, published by China Water & Power Press. The copyright is reserved by National Energy Administration of the People's Republic of China. In the event of any discrepancy in the implementation, the Chinese version shall prevail.

Many thanks go to the staff from the relevant standard development organizations and those who have provided generous assistance in the translation and review process.

For further improvement of the English version, any comments and suggestions are welcome and should be addressed to:

China Renewable Energy Engineering Institute
No. 2 Beixiaojie, Liupukang, Xicheng District, Beijing 100120, China
Website: www.creei.cn

Translating organizations:

POWERCHINA Northwest Engineering Corporation Limited

China Renewable Energy Engineering Institute

Translating staff:

LI Kejia   LYU Kang   YUE Lei

Review panel members:

QIE Chunsheng	Senior English Translator
CHEN Lei	POWERCHINA Zhongnan Engineering Corporation Limited
YAN Wenjun	Army Academy of Armored Forces, PLA
ZHANG Ming	Tsinghua University
GUO Jie	POWERCHINA Beijing Engineering Corporation

QI Wang	POWERCHINA Kunming Engineering Corporation Limited
HUI Xing	POWERCHINA Northwest Engineering Corporation Limited

National Energy Administration of the People's Republic of China

翻译出版说明

本译本为国家能源局委托水电水利规划设计总院按照有关程序和规定，统一组织翻译的能源行业标准英文版系列译本之一。2022 年 5 月 13 日，国家能源局以 2022 年第 4 号公告予以公布。

本译本是根据中国水利水电出版社出版的《光伏发电工程预可行性研究报告编制规程》NB/T 32044—2018 翻译的，著作权归国家能源局所有。在使用过程中，如出现异议，以中文版为准。

本译本在翻译和审核过程中，本标准编制单位及编制组有关成员给予了积极协助。

为不断提高本译本的质量，欢迎使用者提出意见和建议，并反馈给水电水利规划设计总院。

地址：北京市西城区六铺炕北小街 2 号
邮编：100120
网址：www.creei.cn

本译本翻译单位：中国电建集团西北勘测设计研究院有限公司
　　　　　　　　水利水电规划设计总院

本译本翻译人员：李可佳　吕　康　岳　蕾

本译本审核人员：

　　郄春生　英语高级翻译
　　陈　蕾　中国电建集团中南勘测设计研究院有限公司
　　闫文军　中国人民解放军陆军装甲兵学院
　　张　明　清华大学
　　郭　洁　中国电建集团北京勘测设计研究院有限公司
　　漆　望　中国电建集团昆明勘测设计研究院有限公司
　　惠　星　中国电建集团西北勘测设计研究院有限公司

国家能源局

Announcement of National Energy Administration of the People's Republic of China [2018] No. 8

According to the requirements of Document GNJKJ [2009] No. 52, "Notice on Releasing the Energy Sector Standardization Administration Regulations (*tentative*) and detailed implementation rules issued by National Energy Administration of the People's Republic of China", 87 sector standards such as *Wellbore Quality Control Requirements for Coalbed Methane Directional Well*, including 47 energy standards (NB) and 40 electric power standards (DL), are issued by National Energy Administration of the People's Republic of China after due review and approval.

Attachment: Directory of Sector Standards

National Energy Administration of the People's Republic of China

June 6, 2018

Attachment:

Directory of Sector Standards

Serial number	Standard No.	Title	Replaced standard No.	Adopted international standard No.	Approval date	Implementation date
...						
15	NB/T 32044-2018	Specification for Preparation of Pre-feasibility Study Report for Photovoltaic Power Projects			2018-06-06	2018-10-01
...						

Foreword

According to the requirements of Document GNKJ [2013] No. 235 issued by National Energy Administration of the People's Republic of China, "Notice on Releasing the Development and Revision Plan of the First Batch of Energy Sector Standards in 2013", and after extensive investigation and research, summarization of practical experience, and wide solicitation of opinions, the drafting group has prepared this specification.

The main technical contents of this specification include: basic data, project overview, project purpose and scale, solar energy resources, site selection, photovoltaic power generation system design, electrical, general layout, engineering geology and civil works, construction planning, preliminary analysis of environmental impacts, cost estimation, preliminary analysis of financial benefits, and conclusions and recommendations.

National Energy Administration of the People's Republic of China is in charge of the administration of this specification. China Renewable Energy Engineering Institute has proposed this specification and is responsible for its routine management. China Renewable Energy Engineering Institute is responsible for the explanation of specific technical contents. Comments and suggestions in the implementation of this specification should be addressed to:

China Renewable Energy Engineering Institute
No. 2 Beixiaojie, Liupukang, Xicheng District, Beijing 100120, China

Chief development organizations:

China Renewable Energy Engineering Institute

POWERCHINA Northwest Engineering Corporation Limited

Participating development organizations:

POWERCHINA Zhongnan Engineering Corporation Limited

POWERCHINA Beijing Engineering Corporation Limited

HYDROCHINA Corporation Limited

Chief drafting staff:

XIAO Bin	LYU Kang	YANG Mingqun	WU Chengzhi
LIAO E	QIN Xiao	ZHANG Dejian	LYU Song
LIU Huangcheng	ZHANG Yan	WU Laiqun	LIU Xiaoru
MA Gaoxiang	FAN Xiaomiao	WANG Xudong	LYU Hongwei

YUAN Lijie	SUN Lihong	HUANG Zhiwei	SU Fang
WANG Yue	WEN Ziming		

Review panel numbers:

WANG Jixue	QIN Chusheng	TIAN Jingkui	XUE Lianfang
YU Qinggui	WANG Jilin	HUANG Lin	WANG Xiaolan
AN Fucheng	WANG Zhaohui	LYU Zhou'an	ZHANG Wei
ZHANG Yunjie	LIU Xiaoyun	QI Zhicheng	LIU Qigen
DENG Yu	CHEN Xiaokang	XIE Hongwen	WEI Huixiao
CHEN Yuying	LI Fagui	ZHANG Jie	ZHANG Xiaowei
HUANG Hui	LI Shisheng		

Contents

1	**General Provisions**	1
2	**Basic Requirements**	2
3	**Basic Data**	3
4	**Project Overview**	5
5	**Project Purpose and Scale**	6
5.1	Project Necessity	6
5.2	Project Purpose	6
5.3	Project Scale	6
6	**Solar Energy Resources**	7
6.1	Overview of Regional Solar Energy Resources	7
6.2	Preliminary Analysis of Solar Energy Resources	7
6.3	Preliminary Assessment of Solar Energy Resources	7
7	**Site Selection**	8
8	**PV Power Generation System Design**	9
8.1	Selection of PV Module	9
8.2	Selection of PV Array Operating Mode	9
8.3	Selection of Inverter	9
8.4	Layout of PV Array	9
8.5	Estimation of Annual On-Grid Energy	10
9	**Electrical**	11
9.1	Electrical Primary System	11
9.2	Electrical Secondary System	11
10	**General Layout**	12
11	**Engineering Geology and Civil Works**	13
11.1	Engineering Geology and Hydrology	13
11.2	Civil Works	13
12	**Construction Planning**	14
13	**Preliminary Analysis of Environmental Impacts**	15
14	**Cost Estimation**	16
15	**Preliminary Analysis of Financial Benefits**	17
16	**Conclusions and Recommendations**	18
Appendix A	**Contents of Pre-feasibility Study Report**	19
Appendix B	**Cost Estimation Sheets**	21
Appendix C	**Financial Evaluation Sheets**	25
Explanation of Wording in This Specification		36
List of Quoted Standards		37

1 General Provisions

1.0.1　This specification is formulated with a view to standardizing the principles, work content and depth, and technical requirements for the preparation of pre-feasibility study reports for photovoltaic (PV) power projects.

1.0.2　This specification is applicable to the preparation of pre-feasibility study reports for the construction, renovation, and extension of PV power projects.

1.0.3　In addition to this specification, the preparation of pre-feasibility study reports for PV power projects shall comply with other current relevant standards of China.

2 Basic Requirements

2.0.1 The pre-feasibility study report for a PV power project shall be prepared based on the planning and other early-stage study results. The report shall preliminarily analyze the distribution of solar energy resources, check the restrictive factors affecting the project development, propose main technical schemes, make the cost estimation, assess the initial financial benefits, and put forward preliminary conclusions and recommendations on project development.

2.0.2 The preparation of the pre-feasibility study report shall follow the principles of safety, reliability, technical feasibility, practicality, and cost-effectiveness. The use of novel materials, processes, equipment and technologies shall be encouraged after demonstration.

2.0.3 The contents of the pre-feasibility study report should be in accordance with Appendix A of this specification.

3 Basic Data

3.0.1 The development and construction conditions of a PV power project shall be investigated.

3.0.2 The following basic data shall be collected:

1. Study results in the planning stage of the project.
2. Status and development plan of energy resources.
3. Status and development plan of natural resources utilization.
4. Status and development plan of traffic and transportation.
5. Socio-economic status and development plan.
6. Data on ecological environmental protection, soil and water conservation, mineral resources, military installations, cultural relics preservation, etc.
7. Status and development plan of electric power system.
8. Relevant laws and regulations, policies, standards, etc.
9. Data from nearby reference meteorological stations, including:

    1) Basic information of the reference stations, such as the geographical location, observation field elevation, ambient environment, surrounding obstructions, as well as the time and situation of changes in station site, solar radiation observation instrument, and ambient environment since the establishment of the reference stations.

    2) Average air temperature, extreme maximum air temperature, extreme minimum air temperature, maximum air temperature in daytime, minimum air temperature in daytime, and average monthly air temperature in recent 10 consecutive years or more.

    3) Average precipitation and evaporation in recent 10 consecutive years or more.

    4) Maximum depth of frozen ground and thickness of snow in recent 10 consecutive years or more.

    5) Average wind speed, extreme wind speed over years and its occurrence time, and prevailing wind direction in recent 10 consecutive years or more.

    6) Disastrous weather, such as sandstorm, thunderstorm, rainstorm,

hailstorm, and gale, in recent 10 consecutive years or more.

7) Monthly solar irradiation of each year in recent 10 consecutive years or more, and the solar radiation data of at least one complete year that is measured in the same period as the onsite observation station.

10   Solar radiation observation data of the project site.

11   Reanalysis data of solar radiation in the project site area, in the absence of onsite solar observation data.

12   Topographical map of the site area, with a scale of no less than 1 : 50 000.

13   For building-attached photovoltaics (BAPV) or building-integrated photovoltaics (BIPV), the relevant drawings of the buildings and the layout drawings of surrounding buildings (structures).

14   Geological and hydrological data of the site area, such as the regional geological maps and the geological investigation data of the project area.

15   Prices of main equipment, sources and prices of main construction materials, etc.

16   Other data affecting the construction and operation of the project.

3.0.3 The selection of the reference meteorological stations shall comply with the current national standard GB 50797, *Code for Design of Photovoltaic Power Station*.

3.0.4 The site solar radiation should be observed continuously for at least one year at an interval of five minutes, and the observation items shall include the global irradiation, direct irradiation, diffuse irradiation, and air temperature.

4 Project Overview

4.0.1 The geographical location, development purpose and scale, local solar energy resources, general design scheme, grid connection mode, main technical and economic indicators, and comprehensive benefits, etc. of the PV power project shall be described briefly.

4.0.2 The geographical location map of the project site shall be plotted.

5 Project Purpose and Scale

5.1 Project Necessity

The necessity of the PV power project shall be delineated from the perspective of national sustainable development strategy, with consideration of the factors such as national policies on renewable energy sources, local energy mix, power source optimization, current status and development of power system, and environmental protection.

5.2 Project Purpose

5.2.1 The current economic situation and development plan as well as the status of electric power system and its development plan in the project area shall be described.

5.2.2 The project purpose shall be proposed with consideration of the supply-demand conditions of energy resources and the socio-economic development in the project area.

5.3 Project Scale

5.3.1 The project scale shall be proposed based on the regional energy resources, the current status and development plan of electric power system, the effect and requirements of the project on electric power system, and the project development conditions, etc.

5.3.2 A preliminary analysis shall be made of the power consumption market.

6 Solar Energy Resources

6.1 Overview of Regional Solar Energy Resources

The geographical conditions, climate characteristics, and solar energy resources in the project area shall be described, and a distribution diagram of the solar energy resources in the province, autonomous region, or municipality shall be prepared.

6.2 Preliminary Analysis of Solar Energy Resources

6.2.1 The sources of the solar radiation data shall be identified and the rationale of selection shall be analyzed and demonstrated.

6.2.2 The inter-annual and intra-annual variations of global irradiation in the project area shall be analyzed based on the employed solar radiation data, to conclude the inter-annual and intra-annual variation patterns and characteristics of global irradiation over years.

6.2.3 The average annual global irradiation and monthly average global irradiation shall be calculated based on the employed solar radiation data, and a diagram showing the inter-annual and intra-annual variations of global irradiation over years in the project area shall be plotted.

6.3 Preliminary Assessment of Solar Energy Resources

Based on the calculated average annual global irradiation and monthly average global irradiation, a preliminary assessment of solar energy recourses in the site area shall be made in accordance with the current national standards GB/T 31155, *Classification of Solar Energy Resources—Global Radiation*, in which the annual global irradiation classification and its stability evaluation shall be provided.

7 Site Selection

7.0.1 The site selection shall comply with national and local policies, and be coordinated with the relevant development plans.

7.0.2 The project site shall be preliminarily determined according to the site conditions, distribution characteristics of solar energy resources, grid connection and transportation conditions, with due consideration of natural resources utilization, ecological environmental protection, soil and water conservation, flood control, cultural relics preservation, military control, underground minerals, and social stability, etc.

7.0.3 The land use area of the project site and construction scale shall be stated, and the main advantages and possible problems shall be preliminarily assessed.

7.0.4 The conclusions about site selection shall be proposed, the nature of the land shall be stated, the coordinates of inflection points of the site shall be preliminarily determined, and an areal map of the project site shall be plotted.

8 PV Power Generation System Design

8.1 Selection of PV Module

8.1.1 The state-of-the-art of PV module development shall be described briefly.

8.1.2 The types and main technical parameters of the PV modules shall be proposed through techno-economic comparison based on a comprehensive consideration of the characteristics of solar energy resources, incident photon-to-electron conversion efficiency, technical advancement and maturity, market prices, operational reliability, operating mode, natural environment, etc.

8.1.3 The main technical parameters of the PV modules shall be tabulated.

8.2 Selection of PV Array Operating Mode

8.2.1 A preliminary analysis and comparison of different operating modes, including fixed structure, adjustable fixed structure, single-axis tracker, and dual-axis tracker, shall be made in terms of power generation efficiency, operational reliability, project cost, maintenance cost after completion, etc., to propose the operating mode for the PV array.

8.2.2 For the PV arrays of fixed structure, the monthly irradiation on the inclined surfaces at different tilt angles shall be calculated according to the average annual irradiation and monthly average irradiation, the full-year energy output under different tilt angles shall be proposed, and the installation tilt angles shall be proposed after considering the construction conditions and practical modes of the project. For the tracking PV arrays, the installation tilt angles shall be calculated and determined according to the operating mode of the tracking system and the characteristics of solar radiation.

8.3 Selection of Inverter

8.3.1 The state-of-the-art and the performance of inverters shall be described briefly.

8.3.2 The types and main technical parameters of the inverters shall be proposed based on a preliminary analysis of the factors such as the site construction conditions, project operation and maintenance work.

8.3.3 The main technical parameters of the inverters shall be tabulated.

8.4 Layout of PV Array

8.4.1 The series and parallel connection of PV modules shall be designed, and the scheme of the PV module strings shall be proposed based on the initially selected PV modules and inverters.

8.4.2 The spacing between the PV module strings shall be calculated according to the PV array operating mode and the site conditions.

8.4.3 The layout scheme of the PV arrays shall be preliminarily determined.

8.4.4 In the case of a hybrid development of the PV power project with agriculture, fishery, etc., the requirements of agriculture and fishery for sunlight shall be taken into account.

8.5 Estimation of Annual On-Grid Energy

8.5.1 The theoretical annual power output of the PV power project shall be estimated according to the installed capacity, global irradiation, and PV array operating mode.

8.5.2 The overall efficiency of the PV power generation system shall be proposed according to the unavailable solar irradiation, the temperature effect, the influence of dust and snow accumulation on the PV module surfaces, the influence of compatibility of PV modules, and the power losses of lines, inverters and transformers, etc.

8.5.3 The annual degradation rate of the PV power generation system during the operation period shall be proposed according to the performance characteristics of the PV modules.

8.5.4 The yearly on-grid energy, the average annual on-grid energy, and the equivalent full-load hours in the first year and over years during the operation period shall be estimated according to the theoretical annual power output, and the efficiency and annual degradation rate of the PV power generation system.

8.5.5 An estimation table of the yearly on-grid energy shall be prepared, and the histograms of the monthly on-grid energy in the first year of operation and the monthly on-grid energy during the operation period shall be plotted.

9 Electrical

9.1 Electrical Primary System

9.1.1 The grid connection mode and voltage level, number of circuits of outgoing lines, power transmission distance, transmission capacity, and associated transmission and transformation works shall be described according to the installed capacity of the current stage or that of each stage, and the current status and development plan of the electric power system. The location of the step-up substation or switchyard shall be preliminarily determined.

9.1.2 The main electrical connection scheme shall be preliminarily determined.

9.1.3 The power supply scheme for the station service power system shall be preliminarily determined.

9.1.4 The types and main technical parameters of major electrical equipment, including main transformers, power distribution equipment, and reactive power compensation devices, shall be proposed.

9.1.5 The scheme of collection lines shall be proposed.

9.1.6 The general layout scheme for the power distribution equipment of different voltage levels shall be proposed.

9.1.7 A list of main equipment and materials for the electrical primary system shall be provided. The geographic wiring diagram of grid connection system and the main electrical connection diagram of the PV power project, and the layout plan of the electrical equipment in the step-up substation or switchyard shall be plotted.

9.2 Electrical Secondary System

9.2.1 The dispatching management and operating modes, and the automation design principles of the PV power plant shall be proposed.

9.2.2 The monitoring, protection, and communications schemes for the PV power generation system shall be proposed.

9.2.3 The monitoring, protection, and communications schemes for the step-up substation or switchyard shall be proposed.

9.2.4 A list of main equipment and materials for the electrical secondary system and communications system shall be provided.

10 General Layout

10.0.1 The general layout scheme of the PV power project shall be proposed according to the site coverage, natural conditions, topography and geomorphology, site access conditions, etc. The scheme shall cover the step-up substation or switchyard, PV array zone, roads, and other protective facilities, and the general layout plan of the PV power project shall be plotted.

10.0.2 A list of main technical data shall be proposed, including the installed capacity, the total land use area of the project, the land use areas of PV array zone and step-up substation or switchyard, the building area, plantation area and road area.

11 Engineering Geology and Civil Works

11.1 Engineering Geology and Hydrology

11.1.1 The regional geological conditions shall be described and the regional tectonic stability shall be evaluated. The seismic ground motion parameters and seismic basic intensity of the site area shall be proposed in accordance with the current national standard GB 18306, *Seismic Ground Motion Parameters Zonation Map of China*.

11.1.2 The basic geological conditions of the site area shall be described briefly, including the topography and geomorphology, stratigraphy and lithology, geological structure, adverse geological conditions and geological hazards, hydrogeological conditions, and physical and mechanical parameters of rock and soil mass.

11.1.3 The engineering geological conditions of the site area shall be preliminarily evaluated, and the suggestions on the rock and soil mass physical and mechanical parameters and the ground treatment scheme shall be provided.

11.1.4 Hydrological conditions in the site area shall be preliminarily described.

11.2 Civil Works

11.2.1 The functions, scales, and grades of various buildings (structures) shall be proposed according to the design scheme of the PV power project.

11.2.2 The brackets and foundation types of the PV modules as well as the structures and foundation types of other buildings (structures) shall be proposed according to the engineering geological conditions and the project rank, and the preliminary ground treatment scheme shall be provided as required. In the case that the PV power project is combined with agriculture, fishery, etc., the types of supporting structures shall be proposed based on the specific characteristics.

11.2.3 In the windstorm areas, a preliminary protection scheme shall be proposed for the project site, buildings, and equipment.

11.2.4 For the project that might suffer from geological hazards, a preliminary scheme of hazard control shall be proposed accordingly.

11.2.5 A bill of quantities of civil works shall be provided.

12 Construction Planning

12.0.1 The construction conditions, including the geographical location, natural conditions, and site access conditions shall be described.

12.0.2 The construction general layout scheme shall be proposed; the sources of construction materials, electricity and water, and the means of communication shall be recommended; the scheme of site access and the layout of on-site roads shall be determined tentatively; and the construction standard and length of the on-site roads shall be proposed.

12.0.3 The current national and local policies concerning land utilization shall be briefly described, and the construction land use scheme and the coverage area shall be presented.

12.0.4 The technical requirements and methods for the construction and installation of PV module brackets and foundations, main buildings and main equipment shall be presented.

12.0.5 The master construction schedule shall be preliminarily proposed according to the climatic conditions of the project area, project scale, construction time limit, milestone works, etc.

13 Preliminary Analysis of Environmental Impacts

13.0.1 The basis and criterion of environmental impact assessment and the environmental protection objectives shall be described briefly.

13.0.2 The current status of natural environment, ecological environment, and environmental quality in the project area shall be introduced; the conformance and coordination of the project construction and main functional zone planning with the ecological environment protection planning shall be analyzed; and whether the project construction involves ecological redline and environmentally sensitive areas, such as natural reserves, tourist attractions, and source water protection areas, shall be identified.

13.0.3 The main environmental impacts of the project construction shall be preliminarily analyzed and assessed, particularly when the project construction involves the environmentally sensitive area.

13.0.4 The current situation of soil erosion in the project area shall be summarized, and the impact of the PV power project on the soil erosion in the region where the project is located shall be analyzed.

13.0.5 The measures for environmental protection and soil and water conservation shall be preliminarily proposed.

13.0.6 The benefits of energy conservation and emission reduction during the operation period of the PV power project shall be preliminarily calculated.

13.0.7 The possible positive and negative environmental impacts of the project construction shall be described briefly, and the preliminary conclusions and suggestions of environmental impact assessment shall be put forward.

14 Cost Estimation

14.0.1 The cost estimation shall comprise the preparation instruction and the cost estimation sheets.

14.0.2 The preparation instruction shall include the project overview; general principles and basis; principles, basis and results for calculation of base prices; principles and adopted standard rates for preparation of unit prices; principles, basis and rate for preparation of other costs; principles for preparation of contingencies; and principles and basis for preparation of interest during construction.

14.0.3 The project overview shall summarize the project location, natural conditions, installed capacity, construction content, site access conditions, main work quantities, construction period, fund sources and combination, total investment, static investment, and investment per kilowatt.

14.0.4 The general principles and basis shall comprise the relevant standards, specifications and regulations, the adopted price level year during preparation of the cost estimation, and the design results at pre-feasibility study stage.

14.0.5 The principles, basis and results for calculation of base prices shall include the budgetary unit prices for labors, main material sources and their original prices and budgetary prices, and the main equipment prices.

14.0.6 The principles and adopted standard rates for preparation of unit prices shall include the technical measure cost rate, indirect cost rate, profit rate, and tax rate.

14.0.7 The principles, basis and rate for preparation of other costs shall include the land use fees, project overhead, construction supervision fees, consulting service fees, techno-economic review fees, project acceptance fees, production preparation fees, investigation and design fees.

14.0.8 The principles for preparation of contingencies shall include the basic contingency rate and the price contingency rate.

14.0.9 The format and content of the cost estimation sheets should comply with Appendix B of this specification.

15 Preliminary Analysis of Financial Benefits

15.0.1 The installed capacity, annual on-grid energy, construction duration, and financial evaluation period shall be described briefly, and the basis for financial evaluation shall be stated briefly.

15.0.2 The investment plan and financing plan shall be described briefly; the equity capital as a proportion of total investment and of working capital, the credit line, interest rate, grace period, loan term and repayment mode shall be stated.

15.0.3 The value of each cost component shall be stated, and the total cost of the PV power project shall be preliminarily estimated.

15.0.4 Based on the current fiscal and taxation regulations as well as the industry administrative requirements for PV power projects, the calculation methods and parameters for project revenues, taxes, benefits and its distribution shall be stated, the profitability, financial viability, and debt paying ability shall be analyzed, and the financial evaluation indicators shall be proposed.

15.0.5 A financial sensitivity analysis shall be made in terms of investment, on-grid energy, and feed-in tariff variation according to the actual conditions of the project, and the preliminary financial evaluation conclusions shall be provided.

15.0.6 Financial evaluation sheets shall be prepared, and the format and content should be in accordance with Appendix C of this specification.

16 Conclusions and Recommendations

16.0.1 A preliminary conclusion on the feasibility of the PV power project shall be provided from the perspectives of solar energy resources, project construction conditions, grid connection and power consumption, environmental and financial benefits, etc.

16.0.2 Recommendations on and measures for promoting the project implementation shall be provided.

Appendix A   Contents of Pre-feasibility Study Report

1   Project Overview

2   Project Purpose and Scale

2.1   Project Necessity

2.2   Project Purpose

2.3   Project Scale

3   Solar Energy Resources

3.1   Overview of Regional Solar Energy Resources

3.2   Preliminary Analysis of Solar Energy Resources

3.3   Preliminary Assessment of Solar Energy Resources

4   Site Selection

5   PV Power Generation System Design

5.1   Selection of PV Module

5.2   Selection of PV Array Operating Mode

5.3   Selection of Inverter

5.4   Layout of PV Array

5.5   Estimation of Annual On-Grid Energy

6   Electrical

6.1   Electrical Primary System

6.2   Electrical Secondary System

7   General Layout

8   Engineering Geology and Civil Works

8.1   Engineering Geology and Hydrology

8.2   Civil Works

9   Construction Planning

10   Preliminary Analysis of Environmental Impacts

11   Cost Estimation

12 Preliminary Analysis of Financial Benefits
13 Conclusions and Recommendations

Drawings

Drawing 1 Geographical location of the PV power project site

Drawing 2 Areal map of the PV power project site

Drawing 3 General layout of the PV power project

Drawing 4 Geographic connection of the PV power project to the power system

Drawing 5 Main electrical connection of the PV power project

Drawing 6 Layout plan of the electrical equipment for the step-up substation or switchyard

Appendix B Cost Estimation Sheets

B.0.1 The format and content of main techno-economic indicators of the project should comply with Table B.0.1.

Table B.0.1 Main techno-economic indicators

PV power project				Unit cost of PV module	CNY/W_P	
Location				Unit cost of tracking equipment	CNY/set	
Designer				Unit cost of PV module bracket	CNY/t	
Owner				Unit cost of step-up substation or switchyard	CNY/station	
Installed capacity	MW_P		Main quantities	PV module	piece	
AC capacity	MW			Bracket	t	
PV module capacity	W_P/piece			Inverter	set	
Annual average on-grid energy	kWh			Prefabricated transformer	set	
Equivalent full-load hours	h			Earth-rock excavation	m^3	
Static investment	CNY			Concrete	m^3	
Interest during construction	CNY			Rebar	t	
Total investment	CNY			Pile	m or t	
Static investment per kilowatt	CNY/kW_P					
Dynamic investment per kilowatt	CNY/kW_P		Land use area for construction	Permanent land occupation	ha	
Investment per kilowatt hour	CNY/kWh			Temporary land occupation (renting)	ha	
Staffing	person		Total construction period		month	

B.0.2 The format and content of summary sheet of total cost estimation should comply with Table B.0.2.

Table B.0.2 Summary sheet of total cost estimation

No.	Works or cost item	Equipment procurement cost (CNY)	Construction and installation works cost (CNY)	Other costs (CNY)	Total (CNY)	Proportion to total cost (%)

B.0.3 The format and content of cost estimation of equipment and installation works should comply with Table B.0.3.

Table B.0.3 Cost estimation of equipment and installation works

No.	Equipment and specification	Unit	Quantity	Unit price (CNY)			Total (CNY)		
				Equipment	Installation	In which: Installation materials	Equipment	Installation	In which: Installation materials

B.0.4 The format and content of cost estimation of civil works should comply with Table B.0.4.

Table B.0.4 Cost estimation of civil works

No.	Works or cost item	Unit	Quantity	Unit price (CNY)	Total (CNY)

B.0.5 The format and content of estimation of other costs should comply with Table B.0.5.

Table B.0.5 Estimation of other costs

No.	Works or cost item	Unit	Quantity	Rate (%) or unit price (CNY)	Total (CNY)

B.0.6 The format and content of yearly cost estimation should comply with Table B.0.6.

Table B.0.6 Yearly cost estimation (CNY)

No.	Works or cost item	Total cost	Construction period			
			1st year	2nd year	...	nth year

B.0.7 The format and content of calculation sheet of unit prices should comply with Table B.0.7.

Table B.0.7 Calculation sheet of unit prices

Code of norm:					
Construction method:					
No.	Item	Unit	Quantity	Unit price (CNY)	Total (CNY)
I	Direct cost				
(I)	Direct construction cost				
1	Labor cost	CNY			
2	Material cost	CNY			
	...				
3	Machinery expense	CNY			
	...				
(II)	Cost of technical measures	CNY			

Table B.0.7 *(continued)*

Code of norm:					
Construction method:					
No.	Item	Unit	Quantity	Unit price (CNY)	Total (CNY)
II	Indirect cost	CNY			
III	Profit	CNY			
IV	Tax	CNY			
V	Total	CNY			

Appendix C Financial Evaluation Sheets

C.0.1 The format and content of investment plan and fundraising should comply with Table C.0.1.

Table C.0.1 Investment plan and fundraising

No.	Item	Total	Construction period			
			1	2	…	n
1	Total investment (CNY)					
1.1	Construction investment (CNY)					
1.2	Interest during construction (CNY)					
1.3	Working capital (CNY)					
2	Fundraising (CNY)					
2.1	Equity capital (CNY)					
2.1.1	For construction investment (CNY)					
2.1.2	For working capital (CNY)					
2.1.3	For interest during construction (CNY)					
2.2	Debt capital (CNY)					
2.2.1	For construction investment (CNY)					
2.2.2	For working capital (CNY)					
2.2.3	For interest during construction (CNY)					
2.3	Other capital (CNY)					

C.0.2 The format and content of total cost estimation during operation should comply with Table C.0.2.

Table C.0.2 Total cost estimation during operation

No.	Item	Total	Calculation period					
			1	2	3	4	...	n
1	Power generation cost (CNY)							
1.1	Depreciation charge (CNY)							
1.2	Repair cost (CNY)							
1.3	Wages, welfares, etc. (CNY)							
1.4	Insurance premium (CNY)							
1.5	Material cost (CNY)							
1.6	Other expenses (CNY)							
1.7	Interest expense (CNY)							
1.8	Amortization charge (CNY)							
2	Operating cost (CNY)							

C.0.3 The format and content of profits and profits distribution should comply with Table C.0.3.

Table C.0.3 Profits and profits distribution

No.	Item	Total	Calculation period					
			1	2	3	4	...	n
	Installed capacity (MW$_P$)							
	On-grid energy (kWh)							
	Feed-in tariff (CNY/kWh)							
1	Revenue from power sales (CNY)							
1.1	Operating income (CNY)							
1.2	Output VAT (value-added tax) (CNY)							
2	VAT (CNY)							
2.1	VAT payable (CNY)							
2.2	VAT deductible (CNY)							

Table C.0.3 *(continued)*

No.	Item	Total	Calculation period					
			1	2	3	4	...	n
3	Business tax and surcharges (CNY)							
3.1	City maintenance and construction tax (CNY)							
3.2	Additional tax of education (CNY)							
4	Power generation cost (CNY)							
5	Subsidy income (CNY)							
6	Total profit (CNY)							
7	Make up for the losses of the previous year (CNY)							
8	Taxable income (CNY)							
9	Income tax (CNY)							
10	Net profit (CNY)							
11	Extract for statutory surplus reserve (CNY)							
12	Distributable profits for the investors (CNY)							
13	Profit payable (CNY)							
14	Undistributed profit (CNY)							
15	EBIT (earnings before interest and tax) (CNY)							
16	EBITDA (earnings before interest, tax, depreciation, and amortization) (CNY)							

C.0.4 The format and content of loan repayment plan should comply with Table C.0.4.

Table C.0.4 Loan repayment plan

No.	Item	Total	Calculation period					
			1	2	3	4	...	n
1	Loan and loan repayment (CNY)							
1.1	Cumulative principal and interest at the beginning of the year (CNY)							
1.1.1	Principal (CNY)							
1.1.2	Interest during construction (CNY)							
1.2	Loan of the year (CNY)							
1.3	Accrued interest of the year (CNY)							
1.4	Loan repayment of the year (CNY)							
1.4.1	Payment of interest of the year (CNY)							
1.4.2	Repayment of principal of the year (CNY)							
2	Sources of funding for loan repayment (CNY)							
2.1	Profit for repayment (CNY)							
2.2	Depreciation for repayment (CNY)							
2.3	Amortization for repayment (CNY)							
2.4	Interest charged to cost (CNY)							
2.5	Others (CNY)							
	Total (CNY)							
	Interest coverage ratio (%)							
	Debt service coverage ratio (%)							

C.0.5 The format and content of financial plan cash flow statement should comply with Table C.0.5.

Table C.0.5 Financial plan cash flow statement

No.	Item	Total	Calculation period					
			1	2	3	4	...	n
1	Net cash flow from operating activities (CNY)							
1.1	Cash inflow (CNY)							
1.1.1	Operating income (CNY)							
1.1.2	Output VAT (CNY)							
1.1.3	Subsidy income (CNY)							
1.1.4	Other inflows (CNY)							
1.1.5	Working capital recovery (CNY)							
1.2	Cash outflow (CNY)							
1.2.1	Operating cost (CNY)							
1.2.2	Input VAT (CNY)							
1.2.3	Taxes and surcharges (CNY)							
1.2.4	VAT (CNY)							
1.2.5	Income tax (CNY)							
1.2.6	Other outflows (CNY)							
2	Net cash flow from investment activities (CNY)							
2.1	Cash inflow (CNY)							
2.2	Cash outflow (CNY)							
2.2.1	Construction investment (CNY)							
2.2.2	Investment for maintaining operations (CNY)							

Table C.0.5 *(continued)*

No.	Item	Total	Calculation period					
			1	2	3	4	...	n
2.2.3	Working capital (CNY)							
2.2.4	Other outflows (CNY)							
3	Net cash flow from financial activities (CNY)							
3.1	Cash inflow (CNY)							
3.1.1	Equity capital input (CNY)							
3.1.2	Construction investment loan (CNY)							
3.1.3	Working capital loan (CNY)							
3.1.4	Bond (CNY)							
3.1.5	Short-term loan (CNY)							
3.1.6	Other inflows (CNY)							
3.2	Cash outflow (CNY)							
3.2.1	Various interest expenses (CNY)							
3.2.2	Debt principal repayment (CNY)							
3.2.3	Working capital loan principal repayment (CNY)							
3.2.4	Profit payable (dividend distribution) (CNY)							
3.2.5	Other outflows (CNY)							
4	Net cash flow (CNY)							
5	Cumulative surplus fund (CNY)							

C.0.6 The format and content of balance sheet should comply with Table C.0.6.

Table C.0.6 Balance sheet

No.	Item	Total	Calculation period					
			1	2	3	4	...	n
1	Assets (CNY)							
1.1	Total current assets (CNY)							
1.1.1	Current assets (CNY)							
1.1.2	Cumulative surplus fund (CNY)							
1.2	Project under construction (CNY)							
1.3	Net value of fixed assets (CNY)							
1.4	Net intangible and deferred assets (CNY)							
1.5	Assets from VAT deductible (CNY)							
2	Liabilities and owner's equity (CNY)							
2.1	Total current liabilities (CNY)							
2.1.1	Short-term loan (CNY)							
2.1.2	Others (CNY)							
2.2	Construction investment loan (CNY)							
2.3	Working capital loan (CNY)							
2.4	Subtotal of liabilities (CNY)							
2.5	Owner's equity (CNY)							
2.5.1	Equity capital (CNY)							
2.5.2	Capital reserve (CNY)							
2.5.3	Cumulative surplus reserve fund (CNY)							
2.5.4	Cumulative undistributed profit (CNY)							
	Asset-liability ratio (%)							

C.0.7 The format and content of project investment cash flow statement should comply with Table C.0.7.

Table C.0.7 Project investment cash flow statement

No.	Item	Total	Calculation period					
			1	2	3	4	...	n
1	Cash inflow (CNY)							
1.1	Revenue from power sales (CNY)							
1.2	Subsidy income (CNY)							
1.3	Fixed assets residual value recovery (CNY)							
1.4	Working capital recovery (CNY)							
2	Cash outflow (CNY)							
2.1	Construction investment (CNY)							
2.2	Working capital (CNY)							
2.3	Operating cost (CNY)							
2.4	VAT (CNY)							
2.5	Taxes and surcharges (CNY)							
2.6	Investment for maintaining operations (CNY)							
3	Net cash flow before income tax (CNY)							
4	Cumulative net cash flow before income tax (CNY)							
5	Income tax adjustment (CNY)							
6	Net cash flow after income tax (CNY)							
7	Cumulative net cash flow after income tax (CNY)							

C.0.8 The format and content of equity capital cash flow statement should comply with Table C.0.8.

Table C.0.8 Equity capital cash flow statement

No.	Item	Total	Calculation period					
			1	2	3	4	...	n
1	Cash inflow (CNY)							
1.1	Revenue from power sales (CNY)							
1.2	Subsidy income (CNY)							
1.3	Fixed assets residual value recovery (CNY)							
1.4	Working capital recovery (CNY)							
2	Cash outflow (CNY)							
2.1	Equity capital (CNY)							
2.2	Loan principal repayment (CNY)							
2.3	Loan interest payment (CNY)							
2.4	Operating cost (CNY)							
2.5	VAT (CNY)							
2.6	Taxes and surcharges (CNY)							
2.7	Income tax (CNY)							
2.8	Investment for maintaining operations (CNY)							
3	Net cash flow (CNY)							

C.0.9 The format and content of summary sheet of financial indicators should comply with Table C.0.9.

Table C.0.9 Summary sheet of financial indicators

No.	Item	Unit	Quantity
1	Installed capacity	MW_P	
2	On-grid energy	kWh	
3	Total investment (excluding working capital)	CNY	
3.1	Fixed asset investment	CNY	
3.2	Interest during construction	CNY	
4	Working capital	CNY	
5	Feed-in tariff	CNY/ kWh	
6	Total revenue from power sales	CNY	
7	Total expense	CNY	
8	Output VAT	CNY	
9	Subsidy income	CNY	
10	Sale taxes and surcharges	CNY	
11	Total power profits	CNY	
12	FIRR		
12.1	Project investment (before income tax)	%	
12.2	Project investment (after income tax)	%	
12.3	Equity capital	%	
13	ROI	%	
14	Profit and tax to investment	%	
15	ROE	%	
16	Investment payback period (after income tax)	year	
17	Loan term	year	

C.0.10 The format and content of sensitivity analysis should comply with Table C.0.10.

Table C.0.10   Sensitivity analysis

No.	Item		FIRR (%)	
			Project investment	Equity capital
1	Basic scheme			
2	Investment variation	+10 %		
		+5 %		
		−5 %		
		−10 %		
3	Energy output variation	+10 %		
		+5 %		
		−5 %		
		−10 %		
4	Tariff variation	+10 %		
		+5 %		
		−5 %		
		−10 %		

NOTE   The uncertainty factors may be adjusted according to the actual conditions of the project.

Explanation of Wording in This Specification

1. Words used for different degrees of strictness are explained as follows in order to mark the differences in executing the requirements in this specification.

    1) Words denoting a very strict or mandatory requirement: "Must" is used for affirmation; "must not" for negation.

    2) Words denoting a strict requirement under normal conditions: "Shall" is used for affirmation; "shall not" for negation.

    3) Words denoting a permission of a slight choice or an indication of the most suitable choice when conditions permit: "Should" is used for affirmation; "should not" for negation.

    4) "May" is used to express the option available, sometimes with the conditional permit.

2. "Shall meet the requirements of…" or "shall comply with…" is used in this specification to indicate that it is necessary to comply with the requirements stipulated in other relative standards and codes.

List of Quoted Standards

GB 18306, *Seismic Ground Motion Parameters Zonation Map of China*

GB/T 31155, *Classification of Solar Energy Resources—Global Radiation*

GB 50797, *Code for Design of Photovoltaic Power Station*